IGNEOUS ROCKS

By Anna McDougal

Please visit our website, www.enslow.com. For a free color catalog of all our high-quality books, call toll free 1-800-398-2504 or fax 1-877-980-4454.

Cataloging-in-Publication Data
Names: McDougal, Anna.
Title: Igneous rocks / Anna McDougal.
Description: New York : Enslow Publishing, 2024. | Series: Earth's rocks in review | Includes glossary and index.
Identifiers: ISBN 9781978537880 (pbk.) | ISBN 9781978537897 (library bound) | ISBN 9781978537903 (ebook)
Subjects: LCSH: Igneous rocks–Juvenile literature.
Classification: LCC QE461.M425 2024 | DDC 552'.1–dc23

Published in 2024 by
Enslow Publishing
2544 Clinton Street
Buffalo, NY 14224

Copyright © 2024 Enslow Publishing

Portions of this work were originally authored by Kristen Rajczak Nelson and published as *What Are Igneous Rocks?* All new material in this edition authored by Anna McDougal.

Designer: Claire Wrazin
Editor: Caitie McAneney

Photo credits: Cover, p. 1 Artography/Shutterstock.com; series art (title & heading background shape) cddesign.co/Shutterstock.com; series art (dark stone background) Somchai kong/Shutterstock.com; series art (white stone header background) Madredus/Shutterstock.com; series art (light stone background) hlinjue/Shutterstock.com; series art (learn more stone background) MaraZe/Shutterstock.com; p. 5 ImageBank4u/Shutterstock.com; pp. 5, 12, 18, 22 arrows Elina Li/Shutterstock.com; p. 7 VectorMine/Shutterstock.com; p. 9 Florian Nimsdorf/Shutterstock.com; p. 11 Yes058 Montree Nanta/Shutterstock.com; pp. 12, 25 (top) vvoe/Shutterstock.com; p. 13 George P Atkinson/Shutterstock.com; p. 15 DOUGBERRY/iStock; p. 17 npavlov/Shutterstock.com; p. 18 Tyler Boyes/Shutterstock.com; p. 19 Zelenskaya/Shutterstock.com; p. 21 (top) Max Topchii/Shutterstock.com, (bottom) Igor Tichonow/Shutterstock.com; p. 23 www.sandatlas.org/Shutterstock.com; p. 25 (top) Yes058 Montree Nanta/Shutterstock.com, (bottom) Nadezhda Tulatova/Shutterstock.com; p. 27 Gypsy On The Road/Shutterstock.com; p. 29 Kavic.C/Shutterstock.com; p. 30 (arrows) Maksym Drozd/Shutterstock.com.

All rights reserved. No part of this book may be reproduced in any form without permission in writing from the publisher, except by a reviewer.

Printed in the United States of America

Some of the images in this book illustrate individuals who are models. The depictions do not imply actual situations or events.

CPSIA compliance information: Batch #CWENS24: For further information, contact Enslow Publishing at 1-800-398-2504.

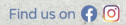

Rocky Earth...4
Molten Magma..6
Hot Lava!..8
Exploring Intrusive Rocks.......................10
Exploring Extrusive Rocks.....................14
Minerals...18
Recognizing Rocks.................................20
Texture..24
Comparing Rocks...................................26
The Rock Cycle.......................................28
Creating Igneous Rocks........................30
Glossary..31
For More Information............................32
Index..32

Words in the glossary appear in **bold** the first time they are used in the text.

ROCKY EARTH

Earth is made up of rocks. The three main kinds are igneous, **sedimentary**, and **metamorphic**. They're formed in different ways. Most of Earth is igneous rock. Igneous rocks start underground as hot, liquid rock called magma. When magma cools and hardens, it changes into rocks we can hold!

lava

volcano

LEARN MORE

Magma is hot, liquid rock under Earth's surface. Lava is hot, liquid rock when it breaks through Earth's surface.

MOLTEN MAGMA

Earth has four layers, or parts: the inner core, outer core, mantle, and crust. Magma is found deep below Earth's surface. It forms in the upper part of the mantle and the lower part of the top layer, the crust. It can reach 1,292° to 2,372°F (700° to 1,300°C)!

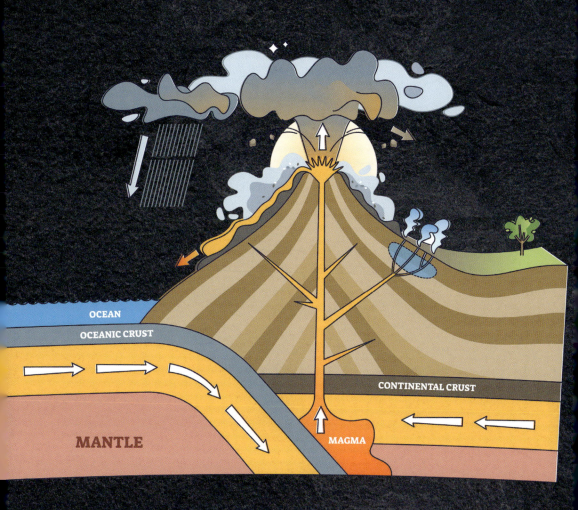

LEARN MORE

Magma includes **molten** rock called the melt, **minerals**, gases, and some solid rock.

HOT LAVA!

The magma that we see on Earth's surface is lava. Most liquid rock that scientists study is lava. It's very hot when it comes out of Earth! This happens during volcanic **eruptions**. When the lava cools, igneous rocks form.

LEARN MORE

Volcanoes are openings in Earth's surface through which lava flows.

EXPLORING INTRUSIVE ROCKS

Two kinds of igneous rock are found on Earth—intrusive and extrusive. The difference is how they form. Intrusive igneous rock forms when magma cools into rock underground. The magma cools slowly, and never reaches Earth's surface.

pegmatite (intrusive igneous rock)

LEARN MORE

Intrusive igneous rock often forms where magma pools in small open spaces or tunnels beneath Earth's surface.

11

When magma cools slowly, igneous rocks have large **crystals**. They can often be seen without a **microscope**. That's one way to tell if a rock is intrusive. Intrusive igneous rocks include granite, pegmatite, and diorite.

diorite

Half Dome granite rock formation at Yosemite National Park

LEARN MORE

Much of the Sierra Nevada mountains are made of granite.

13

EXPLORING EXTRUSIVE ROCKS

What happens when the magma reaches Earth's surface as lava and then cools? The rocks formed in this case are called extrusive igneous rocks. The lava that forms extrusive igneous rock cools very quickly. Extrusive igneous rocks include obsidian, basalt, and pumice.

arrowhead made from obsidian
(extrusive igneous rock)

LEARN MORE

The air **temperature** on Earth is much cooler than the lava, causing it to cool quickly.

Where can you find extrusive igneous rocks? Volcanoes, which look like mountains, have extrusive igneous rocks that build up over time. As lava flows around a volcano, it forms sheets of extrusive igneous rock called a lava field.

LEARN MORE

When a volcano erupts, it may shoot molten rock out of its top. That means extrusive igneous rock can also form far from an active volcano.

lava field

17

MINERALS

Minerals make up all igneous rocks, no matter if they're intrusive or extrusive. If minerals are allowed to "grow," they form crystals. They grow through the slow cooling of hot, liquid rock. That means intrusive rocks have large crystals, while extrusive rocks do not.

olivine

feldspar granite

LEARN MORE

Common minerals in igneous rocks are feldspar and olivine.

RECOGNIZING ROCKS

Intrusive and extrusive igneous rocks look so different because of how they form. Different minerals found in igneous rock also play a part in a rock's appearance, such as its color. For example, rocks rich in iron tend to be darker colors.

LEARN MORE

Igneous rock can be white, pink, gray, black, or brown, depending on the minerals in it.

pink granite (intrusive)

basalt (extrusive)

One way that igneous rocks are grouped is by their silica makeup. Silica is a mineral. Rocks that have the least amount of silica, 45 percent or less, are called ultramafic igneous rocks. Felsic igneous rocks have the most silica at 66 percent or more.

peridotite

LEARN MORE

Ultramafic rocks are low in silica but rich in other minerals, such as olivine and augite. Examples are peridotite and dunite.

TEXTURE

Texture is the size of the crystals or grains in a rock. Extrusive igneous rock tends to be fine, or small, **grained** since it cooled so quickly when it formed. Intrusive igneous rock, which cooled slower, is often coarse grained. That means the grains are bigger.

LEARN MORE

Fine-grained extrusive igneous rock may feel smooth. Coarse-grained intrusive igneous rock may feel bumpy.

coarse-grained granite

fine-grained obsidian

COMPARING ROCKS

Scientists **identify** rocks using color, texture, and other features. Look at these two igneous rocks. The basalt is an extrusive igneous rock. It is dark and looks smooth. The granite is light-colored and coarse-grained. These features help you identify the rocks.

granite

basalt

LEARN MORE

Sometimes basalt looks like a sponge with many little holes in it because of gas bubbles in magma as it erupts.

THE ROCK CYCLE

Igneous rocks are a part of the rock **cycle**. In this cycle, the three main kinds of rocks form, break down, and change over time. Any kind of rock can become molten deep underground. Later, that molten rock cools and hardens as an igneous rock!

ROCK CYCLE

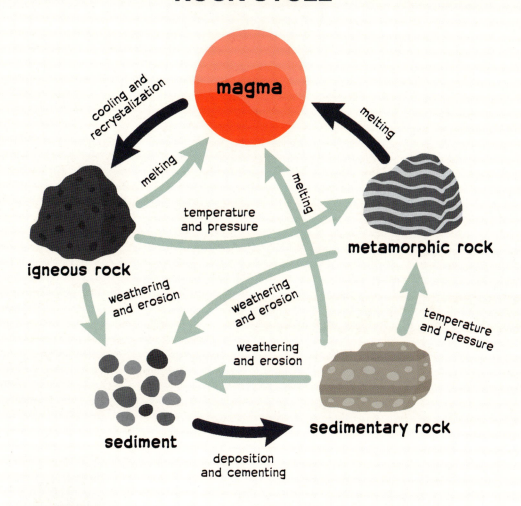

Rocks that break down and are pressed together with other rocks are sedimentary. Rocks that change with temperature and pressure are metamorphic.

CREATING IGNEOUS ROCKS

EXTRUSIVE IGNEOUS ROCKS:

1. Magma reaches Earth's surface.
2 Lava cools.
3. Fine-grained rocks form.

INTRUSIVE IGNEOUS ROCKS:

1. Magma is underground.
2. Magma cools.
3. Coarse-grained rocks form.

crystal: A hard piece of matter formed when something turns into a solid. It often has many sides.

cycle: A sequence of events that repeats.

eruption: The bursting forth of hot, liquid rock from within Earth.

grained: Having to do with the size of the mineral grains that make up a rock.

identify: To find out the name or features of something.

metamorphic: Having to do with rock that has been changed by temperature, pressure, or other natural forces.

microscope: A tool used to view small objects larger or more clearly.

mineral: Matter in the ground that forms rocks.

molten: Changed into a liquid form by heat.

sedimentary: Having to do with rock that forms when sand, stones, and other matter are pressed together over a long time.

temperature: How hot or cold something is.

BOOKS

Owen, Ruth. *Igneous Rocks*. Minneapolis, MN: Bearport Publishing, 2022.

Rogers, Marie. *Exploring Igneous Rocks*. New York, NY: PowerKids Press, 2022.

WEBSITE

Rocks and the Rock Cycle
www.ducksters.com/science/rocks.php
Discover more fun facts about rocks and the rock cycle!

Publisher's note to educators and parents: Our editors have carefully reviewed this website to ensure it is suitable for students. Many websites change frequently, however, and we cannot guarantee that a site's future contents will continue to meet our high standards of quality and educational value. Be advised that students should be closely supervised whenever they access the internet.

crystals, 12, 18, 24

Earth's layers, 6

extrusive, 14, 16, 18, 20, 24, 26, 30

felsic, 22

intrusive, 10, 11, 12, 18, 20, 24, 26, 30

lava, 5, 8, 14, 15, 16

lava field, 16

magma, 4, 6, 7, 8, 10, 11, 12, 14, 27

metamorphic, 4, 29

minerals, 18, 19, 20, 22

rock cycle, 28

sedimentary, 4, 29

silica, 22, 23

ultramafic, 22, 23

volcanoes, 8, 16